CLIMATE CHANGE EDUCATION
Preparing Future and Current Business Leaders

A WORKSHOP SUMMARY

Martin Storksdieck, *Rapporteur*

Steering Committee on Climate Change Education to Prepare Future and Current Business Leaders

Board on Science Education

Division of Behavioral and Social Sciences and Education

NATIONAL RESEARCH COUNCIL
OF THE NATIONAL ACADEMIES

THE NATIONAL ACADEMIES PRESS
Washington, D.C.
www.nap.edu

THE NATIONAL ACADEMIES PRESS   500 Fifth Street, NW   Washington, DC 20001

NOTICE: The project that is the subject of this report was approved by the Governing Board of the National Research Council, whose members are drawn from the councils of the National Academy of Sciences, the National Academy of Engineering, and the Institute of Medicine. The members of the committee responsible for the report were chosen for their special competences and with regard for appropriate balance.

This study was supported by Contract No. DUE-0956031 between the National Academy of Sciences and the National Science Foundation, and also by the Chevron Services Company. Any opinions, findings, conclusions, or recommendations expressed in this publication are those of the author(s) and do not necessarily reflect the views of the organizations or agencies that provided support for the project.

International Standard Book Number-13: 978-0-309-30598-3
International Standard Book Number-10: 0-309-30598-5

Additional copies of this report are available from the National Academies Press, 500 Fifth Street, NW, Keck 360, Washington, DC 20001; (800) 624-6242 or (202) 334-3313; http://www.nap.edu.

Printed in the United States of America

Cover credits: First: Laura DeCapua Photography; second: Mark Bialek; third: Laura DeCapua Photography; fourth: Richard Vietor

Suggested citation: National Research Council. (2014). *Climate Change Education: Preparing Future and Current Business Leaders: A Workshop Summary*. M. Storksdieck, *Rapporteur*. Steering Committee on Climate Change Education to Prepare Future and Current Business Leaders. Board on Science Education, Division of Behavioral and Social Sciences and Education. Washington, DC: The National Academies Press.

THE NATIONAL ACADEMIES

Advisers to the Nation on Science, Engineering, and Medicine

The **National Academy of Sciences** is a private, nonprofit, self-perpetuating society of distinguished scholars engaged in scientific and engineering research, dedicated to the furtherance of science and technology and to their use for the general welfare. Upon the authority of the charter granted to it by the Congress in 1863, the Academy has a mandate that requires it to advise the federal government on scientific and technical matters. Dr. Ralph J. Cicerone is president of the National Academy of Sciences.

The **National Academy of Engineering** was established in 1964, under the charter of the National Academy of Sciences, as a parallel organization of outstanding engineers. It is autonomous in its administration and in the selection of its members, sharing with the National Academy of Sciences the responsibility for advising the federal government. The National Academy of Engineering also sponsors engineering programs aimed at meeting national needs, encourages education and research, and recognizes the superior achievements of engineers. Dr. C. D. Mote, Jr., is president of the National Academy of Engineering.

The **Institute of Medicine** was established in 1970 by the National Academy of Sciences to secure the services of eminent members of appropriate professions in the examination of policy matters pertaining to the health of the public. The Institute acts under the responsibility given to the National Academy of Sciences by its congressional charter to be an adviser to the federal government and, upon its own initiative, to identify issues of medical care, research, and education. Dr. Harvey V. Fineberg is president of the Institute of Medicine.

The **National Research Council** was organized by the National Academy of Sciences in 1916 to associate the broad community of science and technology with the Academy's purposes of furthering knowledge and advising the federal government. Functioning in accordance with general policies determined by the Academy, the Council has become the principal operating agency of both the National Academy of Sciences and the National Academy of Engineering in providing services to the government, the public, and the scientific and engineering communities. The Council is administered jointly by both Academies and the Institute of Medicine. Dr. Ralph J. Cicerone and Dr. C. D. Mote, Jr., are chair and vice chair, respectively, of the National Research Council.

www.national-academies.org

STEERING COMMITTEE ON CLIMATE CHANGE EDUCATION TO PREPARE FUTURE AND CURRENT BUSINESS LEADERS

JANET PEACE (*Chair*), Markets and Business Strategy, C2ES
STEFAN HECK, Sustainability, McKinsey & Company
ANDY HOFFMAN, Erb Institute for Global Sustainable Enterprise at the University of Michigan, Ann Arbor, Stephen M. Ross School of Business
MARK PROEGLER, Climate and Transport Energy Policy, British Petroleum
BRUCE SCHLEIN, Corporate Sustainability, Citigroup
ANANT SUNDARAM, Business Administration, Tuck School of Business at Dartmouth
RICHARD VIETOR, Harvard Graduate School of Business Administration

SHERRIE FORREST, *Study Director* (until July 2012)
CLAUDIA MENGELT, *Study Director* (since August 2012)
MARTIN STORKSDIECK, *Director, Board on Science Education*
ANTHONY BROWN, *Senior Program Assistant* (until August 2013)

Acknowledgments

This workshop summary is based on discussions at a workshop on climate change education for future business leaders, convened on March 14, 2013, by the Board on Science Education of the National Research Council (NRC). The purpose of the workshop was to explore the role that business schools could and should play in preparing future corporate leaders for the challenges and opportunities that climate change will pose. I would like to thank my colleagues who served on the steering committee, each of whom brought deep and varied expertise to the process of planning the workshop. The members of the steering committee developed the agenda, identified and selected presenters and paper authors, and facilitated discussion throughout the workshop. Although they did not participate in writing this report, this summary reflects the insightful comments of the steering committee members, presenters, and many of the workshop participants.

The workshop was generously supported by the National Science Foundation as part of the NRC Roundtable on Climate Change Education, and the Chevron Services Company.

This report has been prepared by the workshop rapporteur as a factual summary of what occurred at the workshop. The planning committee's role was limited to planning and convening the workshop. The views contained in the report are those of individual workshop participants and do not necessarily represent the views of all workshop participants, the planning committee, or NRC.

This workshop summary has been reviewed in draft form by individuals chosen for their diverse perspectives and technical expertise, in accordance with procedures approved by the Report Review Committee of NRC. The purpose of this independent review is to provide candid and critical comments that will assist the institution in making its published report as sound as possible and to ensure that the report meets institutional standards for objectivity, evidence, and responsiveness to the charge. The review comments and draft manuscript remain confidential to protect the integrity of the process. David E. Blockstein, Executive Secretary, Council of Environmental Deans and Directors, and Senior Scientist, National Council for Science and the Environment; Jim Geringer, Director, Esri, Cheyenne, Wyoming; Eugenie C. Scott, Chair of the Advisory Council, National Center for Science Education, Oakland, California; and I were chosen to review the draft report, and I would like to thank my colleagues for volunteering their time to conduct the review in a timely fashion.

Although we as reviewers provided many constructive comments and suggestions, we were not asked to endorse the content of the report nor did we see the final draft of the report before its release. The review of this report was overseen by Cary I. Sneider, Associate Research Professor at the Center for Science Education, Portland State University. Appointed by the NRC, he was responsible for making certain that an independent examination of this report was carried out in accordance with institutional procedures and that all review comments were carefully considered. Responsibility for the final content of this report rests entirely with the author and the institution.

I am grateful for the leadership and support of Martin Storksdieck, director of the Board on Science Education and the Roundtable on Climate Change Education, who also served as the rapporteur for this report. I would like to thank Sherrie Forrest, program officer for the Disasters Roundtable, and Claudia Mengelt, senior program officer for the Ocean Studies Board, for their support in planning and organizing the workshop. Anthony Brown, senior program assistant for the Board on Science Education, deserves our thanks for his invaluable logistical support throughout the project, and Stacee Karras, research associate for the Ocean Studies Board, for her research support. Finally, I would like to acknowledge the support of the Executive Office and the Reports and Communications Office of the NRC Division of Behavioral and Social Sciences and Education. Through their efforts we were able to bring together a diverse group of presenters and participants who shared their experiences and expertise at this workshop. The insights that we gained are summarized in this report. Finally, I would like to extend a special thanks to the speakers and participants whose insightful contributions to the issue of climate change education have been recorded in this workshop summary.

Janet Peace, *Chair*
Steering Committee on Climate Change Education to
Prepare Future and Current Business Leaders

Contents

1

Introduction

Since its establishment in September 2009, the Climate Change Education Roundtable (CCER) at the National Research Council (NRC) has provided a coordinated national network devoted to advancing a dialog about effective, high-quality education addressing the science of climate change and the impacts of climate change on natural, social, and economic systems. The CCER's unique composition brings *ex officio* representatives from the major federal agencies that invest in climate change education together with experts in climate science, the learning sciences, best practices in formal or informal education, and other academic and professional areas relevant to climate change education. (See Appendix A for more information on the CCER.)

Following discussions about the need to address issues around climate change education for future business professionals, the NRC in March 2013 hosted a workshop on "Climate Change Education: Preparing Current and Future Business Leaders" that explored issues associated with teaching climate change-related topics in business schools. The workshop was conducted under the supervision of the Board on Science Education in the NRC's Division of Social and Behavioral Sciences and Education.

The workshop focused on major gaps in our understanding of climate and sustainability education in postsecondary professional schools of business. The workshop connected discourse on climate education for current and future business leaders with the broader discussion on climate change education, in hopes that they influence and benefit each other. The workshop followed a proven format of an NRC-convening activity that allows for decision-relevant exploration of timely and significant topics.

ISSUES AROUND CLIMATE CHANGE EDUCATION FOR CURRENT AND FUTURE BUSINESS LEADERS

Climate change poses challenges as well as opportunities for business and finance and, broadly speaking for the entire economy. Three recent reports by the NRC Committee on America's Climate Choices detail strategies to limit the magnitude of climate change, adapt to the impacts of climate change, and inform an effective response to climate change.[1] These reports make clear how responding fully to the climate challenge will influence the economy of the United States (and the world). Businesses will have to provide services or products with less harmful influence on the climate; respond to a changing policy, regulatory, and market environment; and provide new services and products to help address the challenges that will likely result from climate change (Winn et al., 2011). In the short run, some business and market segments will see these developments as challenging (primarily those responsible for emitting greenhouse gases, such as fossil fuel-based energy production, agriculture, and construction), while others will see business opportunities expand (such as renewable energy production, energy savings technology, and alternative agriculture). The business community, therefore, is divided in its response to current policy proposals to limit greenhouse gas emissions.

Irrespective of the position that businesses take in the public arena, many are beginning to see climate change as another context within which core functions of strategy, finance, operations, marketing, or intersecting with regulatory environments occur, posing risks and opportunities. There are many examples of highly visible activities in which companies pursue opportunities in light of current and future expected changes to the climate system, such as Toyota's marketing of its hybrid Prius and the reinsurance firm Munich Re's operation of a climate intelligence unit. These activities, however, are eclipsed by the myriad of smaller steps the business community can take and is already taking to manage climate risks, and for which the private sector requires capable leaders and managers.

Arguments for why future business leaders need to understand the need to take climate change (or, more broadly speaking, sustainability) into account include operational perspectives (Hoffman, 2005, 2006) as

---

[1]See http://www.americasclimatechoices.org [May 2014] for details on the reports.

well as strategic and moral ones (Blowfield, 2009; Hahn et al., 2010; Patenaude, 2010). Scholars have explored what basic knowledge business leaders need in order to tackle issues related to climate change and global warming (O'Neill Packard and Reinhardt, 2000; Wittneben and Kiyar, 2009). These knowledge needs range from understanding and managing risk and uncertainty (iterative risk management) to system thinking and the ability to listen and communicate to diverse voices and perspectives.

Business schools throughout the United States have responded to this demand. Despite evidence that the issue has not yet penetrated the scholarly field of business and management (Goodall, 2008), the schools have responded in a variety of ways that range from embedding climate-related discussions into courses on risk assessment, leadership, regulatory environments, markets, or other key MBA courses (see Reinhardt and Hyman, 2007, business case 708-26), all the way to offering "green" MBAs or MBAs that focus on issues related to sustainability.

Some MBA programs have "greened" their curriculums not only in an effort to meet the needs of business, but also to respond to the increasing demand from students seeking alternative job paths in sustainability, or who consider climate change a salient issue for their career tracks. As described throughout the workshop, schools are employing myriad approaches to address environmental and sustainability issues in their MBA programs, including (1) incorporating classes on sustainability or the environment into the core curriculum; (2) developing degrees focused on sustainability and related issues; (3) offering specialized electives such as in energy or climate change; (4) creating innovative dual/joint degree programs partnering with other departments or schools; (5) creating teaching materials (cases) or modules for core business school subjects, such as finance, that can be used within the existing core curriculum; or (6) offering extracurricular activities on issues of climate change or sustainability. At Dartmouth College Tuck School of Business, ranked in the top 10 business schools by *U.S. News & World Report* (see footnote 1 for the complete list), for instance, sustainability has been integrated into the core MBA program. In addition, Tuck has developed new programs and courses that focus solely on green issues, including a course on finance within the context of climate change.

In these programs, students can shape their education and career path on issues of sustainability through choosing different portfolios of electives and co-curricular activities. Various universities also offer dual-degree programs between business and law schools (Stanford) or business schools and schools of forestry (Yale) or natural resources (University of Michigan), degree programs that hold the potential to create cross-disciplinary leaders.

The Aspen Institute's Center for Business Education concluded in a

September 2009 report (Shattuck, 2009) that (1) more and more business schools are preparing their students to be environmental and social stewards and ethical actors in the workplace; and (2) in doing so, faculty are drawing on interdisciplinary resources to teach sustainability and instill habits of mind that will serve students well in their careers. Programs and courses on sustainability-related issues prepare graduates and executives to include environmental externalities from ethical, economic, technical, communication, and financial perspectives into strategic business decision making. In addition, courses may focus on technical aspects of climate change adaptation or mitigation, ranging from addressing resource needs in a changing climate (e.g., drinking water availability for bottling plants) to carbon finance and management strategies (CO_2 exchanges, certificates, and permits).

Despite the progress schools have made in incorporating sustainability and green issues into their programs and practices, specifically integrating climate change into curriculum and research agendas has not yet been as prevalent. From 1992 to 1998, only seven articles in the top 30 peer-reviewed management journals included "climate change" or "global warming" in their titles; further, of thousands of courses listed by the Aspen Institute 2010 MBA ranking (Aspen Institute, 2011-2012), only 12 course titles included those two terms (Patenaude, 2011).

Of the top 10 business schools identified in the 2012 *U.S. News & World Report* Best Graduate Schools,[2] three overlapped with a national survey by *The Princeton Review* and *Entrepreneur* magazine that identified the 16 Best Business Schools for Green Business Education[3] (Columbia

---

[2]In order: (1) Stanford University Graduate School of Business; (2) Harvard Business School; (3) MIT Sloan School of Management; (4) University of Pennsylvania Wharton School; (5) Northwestern Kellogg School of Management; (6) University of Chicago Booth School of Business; (7) Dartmouth Tuck School of Business; (8) UC Berkeley Haas School of Business; (9) Columbia Business School, and (10) tied—NYU Stern School of Business and the Yale School of Management. See http://www.usnews.com/education/best-graduate-schools/top-business-schools/articles/2011/03/15/stanford-tops-harvard-in-2012-best-business-schools-rankings [May 2014].

[3]*The Princeton Review* and *Entrepreneur* magazine, 16 Best Business Schools for Green Business Education, in order: (1) Babson College F.W. Olin Graduate School of Business, (2) Bentley University McCallum Graduate School of Business, (3) Brandeis University International Business School, (4) Claremont Graduate University Peter F. Drucker and Masatoshi Ito Graduate School of Management, (5) Clark University Graduate School of Management, (6) Columbia University Columbia Business School, (7) Duke University The Fuqua School of Business, (8) McGill University Desautels Faculty of Management, (9) New York University Leonard N. Stern School of Business, (10) Portland State University School of Business Administration, (11) Stanford University Graduate School of Business, (12) University of Michigan Stephen M. Ross School of Business, (13) University of North Carolina at Chapel Hill Kenan-Flagler Business School, (14) University of Notre Dame Mendoza College of Business, (15) University of Virginia Darden Graduate School of Business; and

Business School, New York University Stern School of Business, and Stanford Graduate School of Business), although most of the other top 10 schools do incorporate some sustainability/environmental issues into their courses and programs. The survey was based on criteria in four key areas: (1) the amount of research the school conducts related to sustainability, (2) availability of courses in sustainability, (3) percentage of faculty teaching such courses, and (4) career services for students interested in green business/social responsibility employment. Below is a brief overview of several programs recognized for their leading programs in this area:

- Stanford has created a dual MBA/MS in Environment and Resources Program between the Business School and the Emmett Interdisciplinary Program in Environment and Resources in the School of Earth Sciences. They offer seminar series in a variety of climate topics as part of their Center for Social Innovation.
- Columbia Business School offers an Individual, Business, and Society (IBS) curriculum as part of the Sanford C. Bernstein & Company. Center for Leadership and Ethics. The center provides students with frameworks and tools to think critically about ethical conflicts and trade-offs, and includes a broad theme of corporate social responsibility. Specific courses, such as "Business in Society" and "Finance and Sustainability," take the approach that profitable financial services can simultaneously create sustainable value for society.
- University of Notre Dame Mendoza College of Business offers courses on greening the supply chain and improving environmental accounting, as well as how to become a leader who uses sustainable enterprise strategies to create business value. A lecture series explores issues, ideas, and trends likely to affect business and society over the next decade and how environmental issues can be incorporated into business practices.

Directly or indirectly, climate change education is beginning to find its way into the curriculum of current and future business leaders. Thus, the time seemed ripe for a workshop to explore the role of climate change education in preparing current and future business leaders, and to build relationships between thought leaders and decision makers in the area of climate and sustainability education in business and finance, and private-sector leaders who hire top business school graduates.

---

(16) University of Wisconsin–Madison Wisconsin School of Business. See http://www.entrepreneur.com/article/219236 [May 2014].

The group of experts assembled by the Board on Science Education to organize this workshop was guided by the following Statement of Task (see Box 1-1).

The one-day workshop addressed the above-stated questions through a sequence of panels and audience discussions. In the first panel, industry representatives discussed the opportunities and barriers that businesses face when addressing climate change. They then reflected on the skills and capabilities that graduates from business schools should bring to their enterprises. A second panel of business school deans responded to these demands by discussing barriers and opportunities to include climate-relevant content in the education of their students. A third panel of business school faculty and staff provided an overview of current best practices for including climate-relevant content in the business school

Box 1
Statement of Task

An ad hoc steering committee will plan and conduct a public workshop to feature invited presentations and discussion from leaders in climate and sustainability education from business schools, including educators, deans, and presidents; for profit and nonprofit business leaders from a variety of sectors with varying exposure to climate risk; human resources and other hiring decision makers; experts in decision support; learning researchers and educators with expertise in the domain of climate, and others to address a range of critical issues such as:

- What kinds of knowledge related to climate change are considered essential, valuable, or useful in business organizations, both from the perspective of businesses and from the perspective of business schools?
- Does formal postsecondary climate education provide advantages to future business leaders beyond the typical climate-related business sectors such as energy, transportation, energy-rich manufacturing, or agriculture? How does the marketplace respond to MBAs with climate-relevant education: is there demand?
- What is current practice in including climate change-related knowledge into the curriculum of future and current business leaders? What models are common in MBA programs? What is the rationale/purpose for including climate change directly in the education of business professionals and future business leaders? What are current strategies for including climate issues in the curriculum and what are the pros and cons of doing so?
- What have been the objectives of such courses and what is the state of evidence on their success/best practices?
- What are challenges and barriers in offering such courses?

curriculum and discussed challenges and solutions from the perspective of practice. A fourth panel of university and industry representatives was then charged to respond to the previous discussions and extract common threats and insights. A final open discussion with all workshop participants concluded the day-long proceedings. A detailed workshop agenda can be found in Appendix B. Appendix C provides a complete list of all registered participants, some of whom followed the discussion via remote access. Biographical sketches of committee members and speakers are featured in Appendix D.

2

Dealing with Climate Change: The Business Perspective

The first panel featured four representatives from energy, chemicals, agriculture, and finance, moderated by the chair of the steering committee, Janet Peace. The panelists were Dave Nagel, executive vice president, BP America Inc.; Michael Parr, senior manager, government affairs, DuPont; Valerie Patrick, sustainability coordinator, Bayer Corporation; and Camilla Seth, executive director, environmental affairs, JP Morgan Chase & Co.

Peace opened the panel by affirming the implications of climate change across industry sectors. She raised fundamental questions with regard to climate change and business:

- What is the fundamental information that businesses need in regard to climate change?
- What are opportunities and risks?
- What capacity to address these issues is expected from new hires?
- What knowledge and skills should newly minted MBAs have?
- What business sectors have the highest need for climate-literate graduates, or does the need apply to all firms and industries?

CLIMATE CHANGE AND ENERGY: PROBLEMS AND OPPORTUNITIES

Dave Nagel noted that energy infuses the entire economy. A somewhat small amount of money is invested in alternative energy, and major investments are made into a variety of projects to tackle climate change in the technical and social arenas. Nagel made four suggestions on how top business schools might address climate change:

1. *Educate business students about the context and drivers of energy.*
   From his perspective in a company that works in the energy sector, students ought to be conversant in the area of climate and energy. Issues related to energy not only impact many areas, but also provide challenges and opportunities locally.[1] Nagel stated that business school students ought to be familiar with basic energy statistics.
2. *Ensure basic science literacy for all business students.*
   In his view, future business leaders need a good layperson's understanding of science to critically follow science debates and interpret relevant data, like probability distributions, that form the basis for strategic decisions on future resource allocations. (This topic was addressed later in the workshop.)
3. *Update operations management and finance courses to include energy issues.*
   He suggested the inclusion of issues related to energy in the core curriculum. Business students need to be ready to improve energy efficiency in their company's operations. They need to understand capital investments into energy efficiency from a broad perspective and full accounting, including the risks involved. (Later in the workshop, Dan Reicher expanded on this point by explaining the need for new financial models and investment tools that are appropriate for innovative projects in the area of energy and climate.)
4. *Provide leadership courses that foster integrated system thinking.*
   Climate change is a long-term issue, with global interdependencies, and local variations and priorities, Nagel asserted. It requires leaders who can "look beyond the numbers" and employ integrated systems thinking. Students need to communicate beyond the confines of their particular area of expertise to address the "bigger strategic picture, but also the risks associated with the overall uncertainty." (See also the later discussion about communication, language, and culture.) For instance, technological advances complicate the argument on when renewables might be competitive on cost alone (Chakravarty et al., 2009).

---

[1]Nagel referred to a January 2013 presentation on BP's Energy Outlook 2030, which provides an overview of these issues. See http://www.bp.com/content/dam/bp/pdf/statistical-review/BP_World_Energy_Outlook_booklet_2013.pdf [May 2014].

Nagel explained that students need to be familiar with the complexity of decisions that are impacted by a combination of market forces and societal and political influences, and that business students need familiarity with issues that touch on the next century, rather than the next quarter's results.

In her presentation, Camilla Seth identified fundamental challenges that climate change poses to all businesses:

- potential disruption to production and consumption,
- increased volatility of energy prices,
- uncertainty in the policy context, and
- challenges related to physical impacts from climate change.

A relatively new challenge, she said, is created by "unburnable carbon": That is, the supply of fossil fuels at competitive costs may no longer present the limiting factor in the extraction and use of hydrocarbons; instead, the capacity of the environment to absorb carbon dioxide might pose the limiting factor. Seth asked how this consideration might become more widespread in business and finance decisions. She said this perspective requires identifying critical leverage points where strategic decision making can be aligned with long-term realities. Seth argued that to understand these issues, future MBAs need to embrace the concept of materiality (i.e., physical or material aspects that influence, or are influenced by, management decisions).

Seth explained that optimal leverage points might require that business leaders embrace nontraditional partnerships and cooperation beyond the typical scope. To succeed in these endeavors, students need "soft skills" to work across sectors and with a diverse set of stakeholders and civil society.

Like Nagel, Seth stressed systems thinking and wondered whether modeling should be part of the standard curriculum of business schools. Investments in energy efficiency, clean energy, and renewables pose risks as well as opportunities, she noted, but financing fossil fuels requires careful management of risks associated with the materiality of the underlying environmental challenges.

Michael Parr noted that DuPont derives one-quarter of its annual revenues of $40 billion from agriculture, using biochemical rather than petrochemical routes to materials and fuels. In the 1990s, DuPont was the first major U.S. company to set goals for public greenhouse gas reduction goals. Parr explained that addressing climate change and energy consumption on the *operations* side was ultimately also good for shareholders.

Parr explained how climate change provides DuPont with operational challenges. For example, a DuPont factory in Mississippi was flooded by Hurricane Katrina, despite a 24-foot protection wall, and the company set up a refugee camp at the location. The incident intensified discussions about "stranded assets" and the need for prudent long-term forecasting in light of climate change when making decisions on infrastructure investments. Business school graduates need to think in these longer time frames, beyond the pressure of "meeting the numbers" monthly or, at best, quarterly.

For DuPont, climate change is not just a strategic issue for long-term planning and cost savings, but also changes important markets. For example, DuPont observes current effects of climate change through shifts in its seed sales. These changes illustrate the need to adapt by developing products for new and emerging needs. In that sense, the markets react with more sensitivity than the political discourse. Climate change, Parr noted, illustrates real operational costs but also significant opportunity, and it provides the backdrop for evolving DuPont's concept of sustainability. A lower carbon future provides business opportunities only for those who think ahead and innovate into new markets; an innovation pipeline can help markets adapt to the new realities of climate change. To capitalize on these new opportunities, new business skills are needed, including analytical capacity or basic science literacy. Parr gave as an example the use of comprehensive life-cycle analysis (LCA). A failure to grasp the full scope of a product's LCA, he explained, means to potentially miss important business opportunities. Because energy is a fundamental driver, Parr said, business school graduates need the basic technical and analytical skills to understand how businesses embed (and potentially optimize) energy-related decisions in their operations and their products.

Valerie Patrick explained that Bayer is mostly removed from the end users of its products, as sales to final consumers represent only 15 percent of its revenue. To her, climate change provides a new lens and way of thinking about business, particularly with a focus on the supply chain. More frequent and stronger storms or unusually hot weather add to the cost of operations, and, particularly in Bayer's crop business, uncertainty and unpredictability of weather can mean loss of revenue and loss of business.

Patrick added a challenge salient for a multinational business: reputation. Her company has a strong position on climate science globally, but the lack of social consensus on the science of climate change in the United States means Bayer treads lightly in the United States in order not to lose business partners.

Patrick pointed to business opportunities in markets where climate change mitigation (i.e., avoiding greenhouse gas emissions) is important, as well as where customers are beginning to address the need for adaptations to help with the consequences of climate change. To capitalize on these opportunities, however, requires the ability to learn as a fundamental organizational capacity. Business school graduates, in her opinion, require the foundational knowledge needed to "wear the climate change goggles." This knowledge includes environmental literacy, management acumen, and the ability to understand business processes and culture or the policy environment. The business structure then needs to be set up to make it easy to employ skilled people in appropriate functions, something that supports strategic execution: That is, the willingness to take action today and engage the resources for tomorrow's benefits and to monitor progress in order to ensure that strategic goals are being met. Her expectations for business school graduates include thought leadership and enterprise management, and the ability to assess the materiality of risk. Patrick explained that it is in the latter area where companies experience the effects of climate change. Future business leaders need to be able to execute change management to align their operations to new conditions, something that requires innovative thinking in making the business case beyond the return on investment, for instance by monetizing as many intangibles such as "biodiversity" as possible for a business decision. Patrick summarized her priorities in training future business leaders in the following way: Leaders need the capacity for system thinking; to be creative and flexible; and to establish a culture of problem-solving inside and outside their company to convene, lead, and activate strategic partnerships with unlikely partners.

CULTIVATING SYSTEMS THINKING

In response to a question about how to best cultivate system thinking, the panel pointed to the importance of first-hand and practical experiences. Also helpful are good cases with implicit models that allow the creation of system maps, which visualize key points of leverage and help people see their roles within the overall system. This may require some level of scientific literacy to manage uncertainty within the context of LCA. Additionally, panelists commented, business leaders need to develop the willingness and ability to listen and talk to a variety of people, particularly those who might disagree with them, in order to expand their own thinking and help address broader needs with limited internal and external opposition.

PREPARATION OF MBAs

Panelists were asked if they are looking for MBAs who have the requisite knowledge and skills or, if not, if they provide relevant training in-house. While the panel members acknowledged some forms of on-the-job professional development, they said companies gravitate toward applicants who bring the requisite basic skill set, like a level of science literacy.

In answering a question on where these ideas fit in the core curriculum of business schools, Nagel said business schools should not move away from the basic core curriculum, but rather bring some of these ideas alive through examples and tools within the core education. Patrick, on the other hand, said she wishes more attention were paid to environmental science aspects that undergird all businesses.

Another question touched on developing skills to prepare future business leaders to solve problems rather than manage dilemmas. As a concrete case, how would one prepare business leaders to make a choice not to locate a business near a vulnerable area? Parr responded that students need to be taught in the core curriculum to think about markets more broadly and to include externalities to see where markets may be going. He referenced assets that cannot be moved and might require expensive adaptation if market forces change, be it for climate change or other reasons. Seth added that students need to learn how to ask the right questions about managing risks and to bring these risks and dilemmas up in appropriate decision points and planning sessions.

3

How Should Business Schools Respond to Climate Change-Related Challenges?

A panel of business school leaders, moderated by Richard Vietor, senior associate dean at the Harvard Graduate School of Business Administration, responded to the challenges from the first panel. The panelists included Valerie Suslow, associate dean of the Ross School of Business, University of Michigan; Robert Hansen, senior associate dean of the Tuck School of Business at Dartmouth College; and Anjani Jain, senior associate dean of the School of Management, Yale University.

CURRENT PRACTICES AND LIMITATIONS

Vietor opened the session by asking about the current practice of leading business schools in addressing climate change and about limitations for doing more. Suslow remarked that many business school initiatives, including those related to climate change, may begin with students who often start grassroots activities that can influence their educational experience. In addition, some faculty engage in relevant research and offer elective courses. Extracurricular activities, an important component of business school life, also provide options for engaging with climate- and business-related topics and issues. Hansen cautioned about being guided too strongly by student demand for addressing social issues or sustainability. Dartmouth offers three electives on climate change and others on issues related to sustainability, but did not change the core curriculum. The 2-year MBA program covers most requirements in the first year, and issues related to climate change are touched on within this core curriculum. He agreed with Suslow that extracurricular activities play a powerful role in providing students with opportunities to think about climate change within the context of businesses. Hansen pointed to a wide variety of important topics that lay claim to being more prominently featured in the business school curriculum, from global health to education. He stressed the opportunity cost for including new ideas: "Everything you do is something else you don't do."

Jain discussed the joint commitment to business and society at the Yale School of Management. Yale has no limit on courses students can take in other areas of the university, and the Yale School of Forestry and Environmental Studies provides an important resource for the business school. This includes a joint degree program, at least 10 elective courses between the business and forestry schools, and the joint development of course cases. At Yale, every core course at the business school is taught by at least two instructors, which increases the capacity at the faculty level for curricular integration and facilitates systems thinking.

Vietor said the Harvard Business School offers no core course on climate change, but six electives cover climate change, and 14 faculty members are engaged in environmental initiatives. Climate change was previously covered in a core course but is not currently.

He then asked whether business schools should be engaged in adaptation only or also address mitigation. Suslow responded that both are needed and that faculty indeed address both. She reported on faculty consensus for a need to broaden the perspective that business students should have on critical issues, and that students are encouraged to cross boundaries to explore them. Business students might normally think more narrowly about the business perspective on climate change, she said, but her school tries to use multidisciplinary action learning to provide students with a broader perspective.

Hansen pointed to the fundamental conceptual frameworks that business students should understand before discussing complex issues like climate change. He stated that these discussions ought to be intellectually rigorous and supported by the right kind of tools. In terms of mitigation, they might include the role of discount rates in determining the time horizon of investments, or considerations of marginal benefits and costs when deciding on optimal control policies. Only when students are provided with rigorous intellectual frameworks for rational discussions should they be asked to think about the right mechanisms for addressing climate change. Hansen stated that mitigation issues are sufficiently covered at Dartmouth, but that there are gaps in addressing adaptation. He remarked that addressing adaptation directly may lead to push-back from students and questioned how to think about adaptation absent rigorous and research-based frameworks. He then questioned the line between teaching to make students think critically about issues and advocacy for particular positions or actions. He cautioned that universities should make students think, not provide them merely with solutions. Jain added that some courses straddle the line between mitigation and adaptation. In terms of advocacy, Jain observed that faculty can and should have their own viewpoints or opinions as long as there is no major institutional bias, but intellectual rigor in addressing issues is their purpose and serves students who are interested in and focused on finding solutions.

OBSTACLES TO ADDRESSING CLIMATE CHANGE IN THE BUSINESS SCHOOL CURRICULUM

Vietor followed up by asking about obstacles to addressing climate change in the curriculum. Suslow said the interdisciplinarity of the issue poses challenges that Yale's team-teaching approach can address. But she reiterated that the real challenge is including climate change perspectives for all students, that is, addressing it in the core curriculum. Suslow said students have considerable power over the course offerings and that donors have some influence as well, but reiterated that student grassroots action and persistent demand can create the momentum that establishes new ideas in the core.

Hanson warned that students may not always be the best judges of what to include in the curriculum and pointed again to the opportunity cost associated with change. Business schools teach foundational concepts and tools, leaving few options for adding new material without deleting others. He said the best option for addressing climate issues in the core curriculum would be via cases that teach foundational concepts within the context of climate change. However, a lack of qualified and interested faculty might form a barrier for teaching specific cases. (As Jacob Park later remarked, "bad things happen when you force faculty to teach something they don't want to teach or are not entirely comfortable or capable of teaching.") Hansen also cautioned that while it might be relatively easy to have practitioners teach certain aspects of the curriculum, top business schools require research-based faculty, although this point was later challenged during a discussion on the balance between academic rigor and practical relevance of concepts and tools taught at business schools. A second barrier for Hansen is the availability of quality, in-depth course material and cases. He questioned the focus on "low-hanging fruit," leaving students without the capacity to tackle knotty issues.

Commenting on the Harvard Business School's policy to send students to other countries in order to set up a company to address globalization and entrepreneurship that may be part of the core curriculum, Hansen asked whether climate change should be treated with the same level of attention (as a cross-cutting theme) in the core. Vietor answered that in his opinion, carbon is a big enough issue to merit inserting into the core. He acknowledged, however, a previous point about the need for more appropriate cases on energy and climate, and capacity within the faculty for teaching them: One case at Harvard that addressed climate change was not liked by students and was dropped from a core course. Jain reiterated that the unit of analysis might be a case or a module, which are easier to include than entire courses. He suggested accounting for curricular coverage of climate-related topics at the level of a lecture or case. Yale currently includes two climate-related cases in the core. At that level, there is less worry about what needs to be given up by faculty.

ISSUES AROUND CURRICULUM REDESIGN

An audience question about the role of recruiters was framed as the need for business schools to show relevant jobs and the demand for particular skills since schools respond to supply and demand. Suslow responded that recruiters are indeed important; they tend to look for critical thinking and analytical skills, then train recruits to their culture and particular job requirements.

Hansen brought up the issue of stand-alone courses on the topic of climate change versus integrating the concept throughout the curriculum. He said he saw higher risks of failure in stand-alone courses and cited a technology course eliminated from the core after it did not perform as expected. However, ethics is now integrated into the core at a more fundamental level. Patrick and Vietor acknowledged that embedding new content into the core is harder and takes longer since it raises the question of how to incorporate new ideas into existing functions, and how to address the intellectual independence of faculty.

Jain explained that curriculum redesign at Yale was difficult, but it engaged faculty. Hansen reiterated that the main idea of business school education is not to have students believe in one particular solution to a business challenge, but to engage them in issues and help them form their own perspectives. While no student should leave a business school without understanding externalities, he said, there is a difference between understanding them and thinking about specific solutions.

A question from the audience challenged the panel to discuss how fundamental assumptions about the way the economic and enterprise system works are reflected in the education of business school students. Suslaw responded that these discussions occur both in and outside the classroom and faculty members are part of these discussions, leaving some students confused, which she interpreted as achieving a broader goal of education. The fact that faculty are free to teach however they wish even in the core courses left Vietor wondering about how to address the issue that climate cases compete with others for the same core case space.

Panelists were asked what they would change if they had the power to do so without any constraints. Suslow suggested bringing in more and diverse perspectives into the business school education. She gave as an example that design schools might bring in business perspectives, while business schools could discuss design where appropriate. She cautioned again, however, that faculty tend to preserve the core education, but might be far more willing to add options in the electives. Jain asked whether it would be useful to attract more STEM (science, technology engineering, and mathematics) graduates into business schools, in part to expand a top talent pool that could shrink should international students find increasingly more attractive MBA programs and business careers in their home countries. Hansen said he would design a course to address so-called great issues in society, even as he acknowledged the difficulties in doing so. And while he agreed that business schools should tap into the talent pool of STEM graduates, he said he also wished to attract liberal arts students for their broad perspective. He cautioned that technical expertise alone might not be sufficient for true future leadership. Jain reiterated that students need a deep appreciation for business fundamentals and Hansen specified the need to teach more about uncertainty and how to manage it: He distinguished risk (not knowing what the probability distribution is) from uncertainty (knowing what the distribution is, but not knowing where one is on it).

Panelists were asked to specify the core capabilities for business school graduates and to reflect how they are represented in the curriculum. Suslaw explained her vision to produce students who have foundational skills, can see the broader picture, can think critically, and can solve problems. She said the emphasis on academics has slipped with the new focus on job placement, and aims at changing the culture back toward increased classroom engagement where faculty members are free to challenge and push students. Jain added that students need to be able to ask the right questions, analyze situations, apply conceptual ideas to create solutions, and keep learning. Hansen articulated a goal to educate students to become the leaders of tomorrow's organizations that schools can be proud of. The abilities to be a leader include envisioning the future and creating it in a responsible way. He said a topic like climate change has to be part of this conversation.

4

Current Efforts and Approaches to Incorporate Climate Change Education into Business Schools

Andy Hoffman, Holcim professor of sustainable enterprise at the University of Michigan, moderated a panel that discussed efforts to include climate change in the education of business school students. He was joined by Anant Sundaram, professor of business administration at the Tuck School of Business; Eric Orts, director of the Initiative for Global Environmental Leadership at Wharton; Jacob Park, associate professor of business strategy and sustainability at Green Mountain College; Dan Reicher, executive director of the Steyer-Taylor Center for Energy Policy and Finance at Stanford University; and Melissa Paschall, director of the Business & Environment Initiative at Harvard Business School.

CURRENT CLIMATE EDUCATION EFFORTS AT TOP BUSINESS SCHOOLS

To answer the question about how top business schools are addressing climate-related issues, Sundaram, working with staff from the National Research Council, examined publicly available information from the websites of a convenience sample of 13 business schools (see Table 4-1):

- Every school addresses the topic in *some* fashion, either directly or under the banner of sustainability or energy.
- A vast majority of schools cover it in a course on environmental sustainability or energy economics (which address a broader set of topics than just climate).

TABLE 4-1 Climate Change Education-Related Activities at 13 Selective U.S. Business Schools

	Climate-Focused Classes	Academic Centers Focused on Climate, Energy, or Sustainability	Climate, Energy, or Sustainability Specializations or Dual-Degree Opportunities
Harvard Business School	—	Yes	—
Yale School of Management	—	Yes	Dual degree
Stanford Graduate School of Business	—	Yes	Dual degree
Columbia Business School	Yes	—	—
Haas School of Business, University of California Berkeley	—	Yes	Yes
University of Chicago, Booth School of Business	—	—	—
Tuck School of Business at Dartmouth College	Yes	Yes	Dual degree
MIT Sloan School of Management	—	Yes	Yes
Kellogg School of Management, Northwestern University	—	—	—
Ross School of Business, University of Michigan	—	Yes	Dual degree
Wharton School, University of Pennsylvania	—	Yes	Yes
Darden School of Business, University of Virginia	—	Yes	Yes
UCLA Anderson School of Management	—	—	Yes
TOTAL	**2**/13	**9**/13[a]	**5**/13 spec **4**/13 dual

[a]None is exclusively climate-focused—all address broader issues of environmental sustainability, with climate as a portion.
SOURCE: Sundaram et al. (2013).

- Two schools (Columbia, Dartmouth) offer (half-semester) courses devoted exclusively to climate change, and both of these courses were taught within the context of finance.
- Topics are typically covered in the field of corporate strategy, finance, economics, or public policy.
- Every school offers a myriad of extracurricular activities and/or student organizations focused on climate, energy, or sustainability.

Many schools provide opportunities for students to specialize in fields related to climate change. For instance, the Wharton School offers majors in Environment & Risk Management, Sustainable Business, and Renewable Energy Management; MIT and University of Virginia offer a Sustainability Certificate; and University of California, Berkeley, provides emphases in Corporate Sustainability, and Energy & Clean Technology. Another mechanism for infusing climate change into the curriculum is dual-degree opportunities (e.g., Columbia, Dartmouth, Michigan, and Stanford). Extracurricular activities for students, research coordination between faculty, and instructional material development (mostly cases) are aided by centers and institutes (e.g., Harvard) that provide an institutional home and visibility to the inclusion of climate change, sustainability, and energy as cross-disciplinary topics at business schools. None of the 13 business schools, however, infuses climate change as a specific topic into the core curriculum.

The findings of this limited landscape study were consistent with Genevieve Patenaude's (2011) conclusions that climate change does not feature in top MBA programs overall, or even in programs that focus on *stewardship*. Patenaude (2010) remarked earlier that "thousands of courses that focus on social and environmental stewardship, offered by MBA programs around the world, are listed in the *Beyond Grey Pinstripes* survey [...] Even among these, only 14 have 'climate change,' 'global warming,' or 'carbon' in their titles, and these are mainly electives that reach only a small fraction of students."

Sundaram provided a closer analysis of the two courses that explicitly address climate change: his own course at the Tuck School of Business at Dartmouth College entitled "Business and Climate Change" and a course at the Columbia Graduate School of Business titled "Carbon Finance." The goals for the Business and Climate Change course include

- developing awareness of climate change and the opportunities/ challenges it presents for leaders of shareholder value-maximizing businesses;
- understanding the emerging "climate economy" and what it means to a company, its industry, and the global economy; and

- developing frameworks to assess the market value consequences of firms' exposure to climate change, their fossil fuel use, carbon footprints and greenhouse gas emissions; and economic analysis tools to assess impacts of likely regulatory responses to climate change.

The goals for the Carbon Finance class at Columbia include

- understanding the scientific issues underlying climate change;
- analyzing which financial tools have been used to date, and their relative effectiveness in combating climate change; and
- evaluating financial tools and strategies that might be used in the future, in the context of a carbon-constrained global economy and national and international policy developments.

Sundaram remarked that the courses resemble traditional finance classes, but cover core concepts in finance within the context of climate change. They provide students with concrete tools in finance, with the driving question of how to increase shareholder value. He then provided a brief analysis of two other courses, one at Harvard and one at the University of Michigan, that include climate change within the larger context of environmental concern, resources, or sustainability. The goal for Harvard Business School's "Environmentally Sustainable Strategy and Operations" course overall is to help students judge company effectiveness by

- understanding or navigating strategic and operational challenges posed by problems like climate change and water shortages, as well as emerging regulatory threats and activist pressures;
- identifying how these issues provide opportunities for competitive advantage by increasing revenues, reducing costs, mitigating risks, and raising rivals' costs; and
- learning how to respond to growing demands from buyers, investors, and activists for greater transparency and accountability for the environmental impacts of their companies' operations and supply chains.

The goal for the University of Michigan Business School and Erb Institute's course "Strategies for Sustainable Development" is to discuss key aspects of business within the context of climate science, and includes

- framing sustainability in the language of business;
- identifying the roles that consumer demand, regulatory compli-

ance, cost of capital, and operational efficiency play in corporate sustainability;
- describing how sustainability has impacted corporate strategies and structures;
- applying mainstream business concepts to sustainability issues as they relate to the corporation; and
- explaining how "being green" fits in with the profit maximization objectives of the business organization.

Sundaram noted these courses all frame climate change or sustainability as a business, rather than a corporate responsibility, issue. Various panelists affirmed the perspective that climate change provides an important context within which concrete management skills and tools can be taught.

Sundaram discussed the relationship between course design and student experience (see Figure 4-1). Individual courses that offer bundled or focused content on one end of the spectrum contrast with infused content across the curriculum. The former provides opportunities for understanding the role of climate change in considerable depth, but invariably focuses on a limited range of business issues (such as finance). The latter

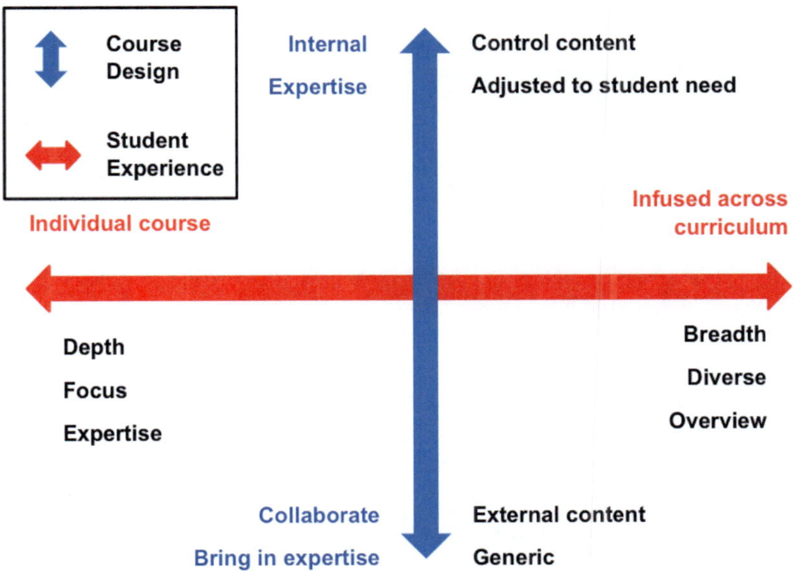

FIGURE 4-1 Relationship between course design and student experience in addressing climate change at business schools.
SOURCE: Sundaram et al. (2013).

allows students to see the implications of climate change across business issues (finance, strategy, operations, etc.), but breadth occurs at the expense of depth and specific capacity development. Expertise in climate science and climate change might be limiting the ability and willingness of faculty to tackle the topic in the classroom. Developing this expertise might be time-consuming or otherwise prohibitive, although it allows focused and highly controlled coverage of climate change within the context of the course. Relevant expertise can be brought in from the outside (e.g., via co-teaching), though content might then be less specific or targeted, and business school instructors might have less control over the content. These two dimensions (infuse across the curriculum versus cover in a separate course, and in-house vs. external expertise) create four quadrants. Which (combination) of those quadrants might best address the needs and specific goals of a particular business school, he explained, will vary by local conditions.

Sundaram closed by reflecting on what seems a common trajectory for addressing cross-cutting concepts in business schools, such as globalization, the digital economy, or business ethics, and he predicted that climate change would go through a similar transition. First, the issue is introduced via elective courses, then becomes infused into the core, where it gets recognized as a stand-alone class, only to get embedded more broadly again across the curriculum (and not offered as a separate core course), perhaps ending up as an expanded set of electives. Sundaram predicted that the idea itself, if salient enough for future business leaders, would ultimately persist in the overall education of business school students if supported by a strong conceptual framework, rapid development of high-quality classroom resources (including cases), and original research. At the moment, he said, too much data around the intersection of climate change and business strategies are missing or inaccessible (proprietary), slowing down current research and therefore the stream of scholarship needed to support the teaching of climate change-related issues in many business schools.

Sundaram noted in this context that business schools are about wealth creation, and they can address two levers of climate change: the energy content of the economy and the carbon content of energy. Business schools cannot (and therefore should not attempt to) tackle other important components of the greenhouse gas emission issue, such as population growth. He said that three key strategies form a conceptual framework around climate change and business: (1) burn fossil fuels more efficiently or more carbon productively, (2) shift to nonfossil fuel sources, or (3) capture carbon dioxide before it is emitted and "put it away for good." He mentioned that this framework requires a functioning global market of carbon dioxide emissions.

Melissa Paschall described the 2.5-year-old Business and Environment Initiative at the Harvard Business School (HBS) as a relatively new player in the arena of climate change and sustainability research and education of a general business school. Organized in a similar way as the established Global Initiative, Entrepreneurship Initiative, and Leadership Initiative, it received instant recognition within HBS.

The initiative focuses on developing and making available free online resources, such as environment-oriented cases, or entire course modules for 4-6 weeks of course material. The initiative has embedded around 15 cases across the required curriculum, some specific to climate, but others with a broader environmental focus. Paschall noted that getting environment-oriented cases into required courses was only part of the challenge: keeping them there is equally difficult, due to overall changes in the mix of cases and the need for interested and capable instructors. In addition to inserting relevant cases into required and elective courses, the initiative works toward embedding climate or environmental aspects into field-based experiences. A student advisory group provides helpful intelligence on current teaching practice for environmental cases and suggests new opportunities for inserting relevant cases or modules into the curriculum. Aside from influencing the curriculum, the initiative is engaged in extracurricular activities and provides the institutional hub for a community of practice around teaching and research on environment-related topics.

Eric Orts described the Wharton School's Initiative for Global Environmental Leadership, which originated with an internal competitive grant from the provost. The initiative now sustains itself with funding from private sources, in part from a corporate community that understands the relevance of sustainability to their businesses. Orts underscored a previous comment that the private sector and students might be more attuned to sustainability questions than academe. The initiative focuses on curriculum change, research, and outreach. Through outreach, business representatives and academics come together to identify "big issues" in the intersection of business and environment. He argued that success for such initiatives depends on partnerships within the university and support from the university's leadership, but that an engaged student body is equally important.

Dan Reicher described the Center for Energy Policy and Finance—a collaboration between Stanford's business and law schools. It tries to emulate the creativity in technology development by developing new policy and finance tools for funding energy projects, tools that bridge the "valley of death" between early development and commercialization. Aside from overseeing research, Reicher also co-teaches a course on finance with climate change as a significant element. The center works across the entire university on extracurricular issues (e.g., a presentation by the two pilots who first flew a solar-powered plane across the United States). Reicher argued that initiatives like his need to fit in with the predominant culture of their host universities.

Jacob Park corroborated Reichert's comment about institutional fit. He said the comparative advantage of teaching climate change at a small liberal arts college lies in the "high touch" (small student/teacher ratio) and the ease of collaborating with colleagues. Park explained that community-based service learning, where students solve problems at local businesses, government agencies, or nonprofits, allows for field-based work that is innovative, relevant, and not too expensive, and, while not unique to his school, is far easier to implement at a small liberal arts college than within the context of a major business school. Aside from cost, he identified sufficient numbers of appropriate faculty as one of the main barriers, but also reflected on the tension or trade-off between academic rigor and societal relevance. Park questioned whether, in pursuit of rigorous peer-reviewed journal articles, academics are creating relevant work that serves students.

Andy Hoffman explained that the University of Michigan's Erb Institute offers a dual-degree program where students could earn an MBA and a Masters in Natural Resources within three years. The institute itself provides a sound institutional support structure through its considerable endowment and a faculty champion. Hoffman stated that it would be "deadly" to frame the relationship between climate change and business as corporate social responsibility. Instead, climate change and sustainability are taught at his university as shifting markets, consumer demand, operational efficiency, compliance, etc., in other words, core concepts that are at the heart of a business school education. Hoffman agreed with previous panelists that students are a force for change. He said the possibilities for cross-disciplinarity are enormous. As an example, his course in green construction was listed in three different schools within the university. He also stressed the benefit of working with external science partners to ensure that data on the "materiality" of doing business can be used effectively in making business decisions. Hoffman also reiterated the need to challenge students while providing a balanced perspective, in part by bringing in outside speakers who represent the spectrum from conservative to liberal on climate change, and he wondered whether the question of core versus electives should be amended by also asking where innovation might most likely be coming from.

WHAT WORKS, AND WHAT DOES NOT?

Hoffman summarized four options for addressing climate change in the business school curriculum: (1) offer a course in the core curriculum, (2) provide elective courses, (3) infuse climate-relevant ideas throughout the curriculum (e.g., via cases), and/or (4) provide a range of extracurricular activities. He then challenged the panel to advise a faculty member on key concepts to provide business students with climate change-relevant skills and knowledge. The panel generated the following ideas:

- Find other allies on the faculty, whether in law and ethics, economics (externalities, discount rates), operations, supply-side management, or finance.
- Find solid financial support, whether through the dean, outside funding, or donors and benefactors.
- Ensure sustained student interest and focus any course on providing students with tools and frameworks for better managerial decisions.
- Think about a sideways approach if appropriate, for instance, in a "clean technology course" where climate change is not addressed directly, but as part of a broader question on "how to make money addressing this by approaching it in a politically viable way."
- Include "real world" people who work on these issues and have them interact with students.
- Connect the issue of climate change with the core identity of the school, such as innovation in Silicon Valley, where Stanford in located.
- Affirm the need to offer something to students. Once affirmed, ask how it can be accomplished and what resources are needed.
- Get a basic education in climate science.
- Do not require faculty to teach potentially controversial issues if they do not want to or feel that they cannot teach such a topic.
- Build excitement about the saliency of the issue and be passionate about the issue: Students get it when teachers are passionate and inspiring.
- Make it seem important and not peripheral. Student interest can decline if a course, program, or school indicates that the issue is not considered important or essential.

CLIMATE CHANGE AS PART OF BROADER SUSTAINABILITY, OR AS A SEPARATE TOPIC?

Hoffman asked the panel about the pros and cons of addressing climate change directly versus under a broader framework, such as sustainability, or a recognized issue, such as energy. The panel offered the following thoughts:

- It depends on what the possible taboo issues might be, where you are, and what you represent. Climate change can be addressed indirectly through a range of topics, from emerging markets to operational efficiency, and can get connected to business opportunities.
- As long as climate change is not perceived as value-neutral (in the United States), it may be problematic to address it directly in many business school contexts. A sustainability frame might be more productive in many contexts, although "sustainability" itself might be considered controversial and difficult to conceptualize.
- The perspective should be focused on finding solutions to problems, rather than discussing the problem itself, and ensuring that the framing resonates with students.
- The difference with the European (or international) perspective, where climate change is seen as part of sustainability and where other environmental topics also receive considerable attention, should be kept in mind. In the spirit of systems thinking, climate change can serve as the gateway into other global issues, like water availability, and vice versa.

Responding to a workshop participant's question, the panel speculated that interest in climate and sustainability might be highest among undergraduates and lowest in participants who attend executive programs. MBA students may wonder about the payoff in learning about climate change, and PhD students might worry about their professional identity and their chances on the academic job market.

A workshop participant wondered how faculty can help students go from recognizing the problem to empowering them to work towards solutions. Panel members said students could get excited about becoming part of the solution while being successful in business. For younger students, climate change and sustainability will be a far more salient issue over the course of their careers. Already 50-60 percent of Standard & Poor's-listed businesses employ a chief sustainability officer. However, the issue was raised whether practitioners who are teaching practice-oriented courses at universities that can sometimes be quite exciting to students are tapping sufficiently into the scholarship published in academic journals, and how to bridge the gap between research/theory and the practice entrepreneurs.

Various workshop participants asked how to bridge between academic silos within business schools (or academia in general). Paschall agreed that it is difficult not to be siloed, but centers and initiatives can serve as conduits and incubators for collaboration and cooperation. Reicher affirmed that this occurs at Stanford, where cross-disciplinary work ensures access to needed expertise. He also offered a different perspective on silos as obstacles for tackling broader issues such as climate change. He said climate change might serve as a means to finally address academic issues around siloing or the debate between rigor and relevance (i.e., the question of well-researched theoretical frameworks versus using ad-hoc solutions to address real problems and challenges).

Another endemic problem in academia that can stand in the way of addressing climate change more broadly in education and research at business schools is the process for promotion and tenure, Reicher added. Orts offered that focusing on controversial issues or trying to change university culture might best be reserved for post-tenure, and that junior faculty has to consider the predominant culture while pursuing what they feel passionate about. He suggested that junior faculty find like-minded faculty members, think about connections to mainstream areas of business, find ways to publish in respected (high-impact) journals, and think about quality scholarship while focusing on salient issues.

A question about basic science literacy led to divergent perspectives from the panel. Some called for basic understanding in order to locate business decisions within the context of climate change, while others pointed out that a more in-depth understanding of climate science might not be necessary, particularly when this expertise and knowledge can be brought in through collaboration and cooperation. For example, Hoffman wondered whether a company that produces electric vehicles needs to understand climate change or would be better off with a deep appreciation of market fundamentals to succeed. If he had to choose, he said he would prefer the latter even as he understood the importance of the former. Hoffman then reflected on the use of the term "climate change" within the business school context and provided an analogy from social change research. He stated that when the gay community was "in the closet," no social change toward stronger equality and inclusion could occur. He said that while at times unpleasant, retreating from the term would mean postponing the time when the issue will become mainstreamed within U.S. business schools. He suggested combining discussion about the science of climate change with discussion about what the business community and individual businesses could do about it. Sundaram offered that the science behind climate change is important, but uncertainty persists about when to take what action. Yet, finance and economic research deal with issues of risk and uncertainty all the time. Ultimately, he said, climate-related questions fit very well into a business school curriculum that teaches students how to grapple with uncertainty and risk.

5

Needs and Opportunities: Toward a Way Forward

Workshop committee member Anant Sundaram moderated a final panel of experts from business and academia who reflected on the previous discussions. The panel included Dan Vermeer, executive director of the Center for Energy, Development and the Global Environment at Duke University; Laura Clise, director of sustainable development at AREVA Inc.; and Bruce Schlein, director of corporate sustainability at Citi.

Vermeer challenged the group to think about climate change not as a question of responsibility or sustainability, but as fundamental to almost every aspect of a business operation, and hence part of the worldview and repertoire of every business professional. He reminded participants that the world population is not only projected to reach 9 billion soon, but also the share of the population that will aspire to a middle-class lifestyle with its associated energy consumption and environmental impact is projected to grow even more. The likely transformations that are needed to serve this growing number have implications not only for the environment, but also for overall infrastructure, creating new market environments in which future business school graduates will operate. He noted that he heard similar messages across presenters and panels. Business leaders discussed the need for systems thinking, integration of the scientific and social dimensions of global issues, breaking out of silos, development of a growing interest and ability to anticipate and internalize cost, focus on operational risks and costs, and the importance of basic science literacy. University leaders discussed avenues to address climate change in business schools, from stand-alone courses to infusing ideas across the entire curriculum and into cases, exercises, lectures, or field experiences. He was reminded of the many options to infiltrate the curriculum in more imaginary ways than another elective or one slot in the core curriculum, including extra- and co-curricular activities and experiential learning.

To Vermeer, the goals for a business school education are shifting and expanding, from a historic focus on core business functions to a broader vision of competencies, including cross-cultural communication, and working across functions and disciplines. He noted that an increasing number of students want to harness the market to create value in sustainable ways, and many students see emerging markets as places for future growth and opportunity. He suggested a role for business schools to present sustainability as a lens for making use of these new opportunities.

Laura Clise started her remarks with two haikus (see Box 5-1).

Clise also stressed the importance of framing climate change as a business issue. She urged a focus on opportunities and solutions that might begin by cultivating excitement from high school to MBA students about how addressing climate change might present career opportunities. She also supported the perspective that climate change could provide a frame for talking about fundamental aspects of business operations like entrepreneurship and innovation, with climate change as the context within which to prepare tomorrow's organizational and business leaders. She said a new generation of students wants to address big issues by starting something new or by changing existing organizations. Despite coming from an engineering-oriented firm, she questioned whether scientific literacy is necessary, or whether being "multilingual" to develop language that helps bridge cross-cultural differences is more fundamental to success.

BOX 1
Laura Clise's Haiku Poems

Haiku 1:
Relevant today
Climate change education
A business issue

Haiku 2:
Climate change presents
Risk and opportunity
Call it what it is

Bruce Schlein prefaced his remarks by saying that his three-person team on sustainability relies on others in their organization to effect change. He reflected on the qualities his colleagues need and how these qualities translate into a business school curriculum. Schlein said they need to be a "jack of all trades and a master of some." The ability to be a "master of some" should come from existing core curricula, but more general skills ("jack") could be obtained outside of the core curriculum or even outside the school itself. Schlein's biggest question was the optimal balance of the skills necessary for the intersection between generalist and expert. Those include systems thinking and systems doing, bridging different dimensions (within firms and their silos of expertise, with civic society and policy, and internationally), and talking to people with different viewpoints. Schlein noted that not everyone has the ability to acquire these diverse skills. He argued, though, that a process ought to be embedded into the core curriculum for honing the basic skills needed to find the balance, and that topical issues like global health or climate change could then be plugged into this process.

Schlein noted that the expertise to function effectively in the intersection of generalist and expert is too often dismissed as a soft skill, but should be defined and articulated as an essential core skill. Lastly, students need to develop a capacity for innovation and disruption. He explained that Citi, like most other banks, is organized by property asset class lines. Energy issues traverse these lines and are therefore disruptive changes that require new organizational structure entrepreneurs.

Sundaram observed that the first panel described the basic skills of a business school graduate: see new opportunities, understand associated risks, and lead the necessary changes or actions. In the context of climate change, they entail longer-term thinking, recognizing and internalizing externalities, appreciating the materiality of business ventures, embracing nontraditional partnerships, developing "soft skills" to work across industries and sectors, and systems thinking and modeling. Sundaram described his take-away messages from the "deans' panel": Climate change might best be embedded into the curriculum, electives can complement a busy core curriculum, the quality of research needs to be improved, and business schools should be vigilant about institutional bias, even as individual faculty express strong personal views.

During a subsequent audience discussion, the issue of new standards for elementary and secondary science education was mentioned (National Research Council, 2011b, 2012a, 2013). Their new orientation toward scientific practices, and the inclusion of climate change and climate science, might mean that in the future, more students with sufficient basic preparation in the natural and physical sciences enter business schools.

Participants noted that climate change has become the context within which students need to learn and practice core business skills, but that, despite some initial progress presented during the workshop, much work is still needed to prepare the majority of students in that way.

Dan Reicher used a metaphor of getting sick from polluted drinking water: Contamination could occur at the well, during transportation, storage, or consumption. The same is true for climate change, he asserted. Lack of preparation or utilization could occur during the entire "supply chain" of education. Reichert said that he sees no other choice but to embed the issue into the core education of future business leaders. Sundaram ended the panel discussion by noting changes over the last 10 years, when the topic itself would not have been on the horizon for business schools; even five years ago, the conversations were not substantive. Now, he said, business schools are beginning to catch up to where many businesses already are.

Chair of the workshop organizing committee Janet Peace and Martin Storksdieck, director of the Board on Science Education at the National Research Council, ended the workshop with closing remarks. Peace again noted the key issues raised as competencies that graduates should acquire or hone at business school, including systems thinking, the ability to talk to others (in terms of language and discipline), cross-cultural competency to work in diverse teams, an ability to bridge to science professionals, and the capacity to take the long view. She said business schools should teach critical thinking skills within the context of addressing climate change as a business issue.

Storksdieck reiterated the need for business schools to support the development of a person who is a "jack of all trades and master of some" and something in between. He referred to a recent National Academies' report, *Education for Life and Work* (National Research Council, 2012b), which identifies 21st-century skills as increasingly more foundational for success in life. He noted that the workshop discussion showed how business leaders need these skills and abilities, and that core pedagogical concepts from K-12 science education, like learning through relevant practices, are also embraced by an increasing number of business schools. He noted that business schools are good at defining visions and inputs/resources when discussing educational attainment, but concrete learning outcomes with measurable metrics for immediate impact are still missing. Storksdieck also noted the difference between skills and abilities needed to secure the first job and those needed to succeed long term. Are they the same, he queried, and how do business schools decide what to focus on if they are not?

Storksdieck closed by comparing climate change education with improvements to undergraduate science education. While improvements to traditional instruction have been established through research, implementing them more broadly requires a broader systemic approach toward change. The same patient and systemic perspective might be needed to improve climate change education at U.S. business schools.

References

The Aspen Institute. (2011-2012). *Beyond Grey Pinstripes.* Available: http://www.beyondgreypinstripes.org/rankings [February 2014].

Barrett, S. (2009). The coming global climate-technology revolution. *Journal of Economic Perspectives, 23*(2), 53-57.

Beinhocker, E., Oppenheim, J., Irons, B., Lahti, M., Farrell, D., Nyquist, S., Remes, J., Nauclér, T., and Enkvist, P. (2008). *The Carbon Productivity Challenge: Curbing Climate Change and Sustaining Economic Growth.* Report for McKinsey Global Institute. Available: http://www.mckinsey.com/insights/energy_resources_materials/the_carbon_productivity_challenge [February 2014].

Blowfield, M. (2009). Global warning. *Business Strategy Review, 20*(2), 74-79.

British Broadcasting Corporation. (n.d.). Climate change. Available: http://www.bbc.co.uk/search/climate_change [February 2014].

Carbon Disclosure Project. (2012a). *CDP S&P 500 Report 2011: Strategic Advantage Through Climate Change Action.* Available: https://www.cdp.net/CDPResults/CDP-2011-SP500.pdf [February 2014].

Carbon Disclosure Project. (2012b). *CDP Global 500 Report 2011: Accelerating Low Carbon Growth.* Available: https://www.cdp.net/CDPResults/CDP-G500-2011-Report.pdf [February 2014].

Carbon Tracker Initiative. (2013). *Unburnable Carbon: Are the World's Financial Markets Carrying a Carbon Bubble?* Available: http://www.carbontracker.org/wp-content/uploads/downloads/2011/07/Unburnable-Carbon-Full-rev2.pdf [February 2014].

Chakravarty, S., Chikkatur, A., de Coninck, H., Pacala, S., Socolow, R., and Tavoni, M. (2009). Sharing global CO_2 emission reductions among one billion high emitters. *Proceedings of the National Academy of Sciences, 106*(29), 11,884-11,888.

Congressional Budget Office. (2008). *Policy Options for Reducing CO_2 Emissions.* Available: http://www.cbo.gov/sites/default/files/cbofiles/ftpdocs/89xx/doc8934/02-12-carbon.pdf [February 2014].

Daley, J. (2011). Top business schools for eco-enterpreneurs. *Enterpreneur.* Available: http://www.entrepreneur.com/article/219236 [May 2014].

Darbee, P.A., and Field, C.B. (2010). *Climate Change for Policymakers and Business Leaders*. San Francisco, CA: Pacific Gas and Electric Company. Available: http://energyseminar.stanford.edu/sites/all/files/eventimage/PGE%20climange%20change%20for%20policymakers%20and%20business%20leaders.pdf [February 2014].

Enkvist, P., Nauclér, T., and Oppenheim, J.M. (2008). Business strategies for climate change. *The McKinsey Quarterly*, 24-33. Available: http://sites.middlebury.edu/ec465/files/2011/02/Business-strategies-for-climate-change.pdf [February 2014].

Financial Times. (2014). Climate change. Available: http://www.ft.com/intl/indepth/climate change [February 2014].

Friedman, T.L. (2009). *Hot, Flat and Crowded: Why We Need a Green Revolution—and How It Can Renew America*. New York: Farrar, Straus, and Giroux.

Goodall, A.H. (2008). Why have the leading journals in management (and other social sciences) failed to respond to climate change? *Journal of Management Inquiry, 17*(4), 408-420.

Grantham, J. (2010). *Everything you need to know about global warming in five minutes*. Boston: GMO. Available: https://www.gmo.com/America/MyHome/ [February 2014].

Greenbiz. (2010). All blogs. Available: http://www.greenbiz.com/engage/blogs [February 2014].

Hahn, T., Kolk, A., and Winn, M. (2010). A new future for business? Rethinking management theory and business strategy. *Business and Society, 49*(3), 385-401.

Hoffman, A.J. (2005). Climate change strategy: The business logic behind voluntary greenhouse gas reductions. *California Management Review, 47*(3), 21-46.

Hoffman, A.J. (2006). *Getting Ahead of the Curve: Corporate Strategies That Address Climate Change*. Report for the Center for Climate and Energy Solutions. Available: http://www.c2es.org/docUploads/PEW_CorpStrategies.pdf [February 2014].

Hoffman, A., and Woody, J.G. (2008). *Climate Change: What's Your Business Strategy? (Memo to the CEO)*. Cambridge, MA: Harvard Business School Press.

Hone, D. (n.d.). Shell blog. Available: http://blogs.shell.com/climatechange/ [February 2014].

HSBC Global Research. (2010). *Sizing the Climate Economy*. Available: http://firstforsustainability.org/media/HSBC%20Sizing%20the%20Climate%20Economy%20Sept%202010.pdf [February 2014].

Intergovernmental Panel on Climate Change. (2007). Summary for policymakers. In S. Solomon, D. Qin, M. Manning, Z. Chen, M. Marquis, K.B. Averyt, M. Tignor, and H.L. Miller (Eds.), *Climate Change 2007: The Physical Science Basis. Contribution of Working Group I to the Fourth Assessment Report of the Intergovernmental Panel on Climate Change*. Available: http://www.ipcc.ch/pdf/assessment-report/ar4/wg1/ar4-wg1-spm.pdf [February 2014].

Metcalf, G.E. (2009). Market-based policy options to control U.S. greenhouse gas emissions. *Journal of Economic Perspectives, 23*(2), 5-27.

National Research Council. (2011a). *Climate Change Education: Goals, Audiences, and Strategies: A Workshop Summary*. S. Forrest and M.A. Feder, Rapporteurs. Board on Science Education, Division of Behavioral and Social Sciences and Education. Washington, DC: The National Academies Press.

National Research Council. (2011b). *A Framework for K-12 Science Education: Practices, Crosscutting Concepts, and Core Ideas*. Committee on a Conceptual Framework for New K-12 Science Education Standards. Board on Science Education, Division of Behavioral and Social Sciences and Education. Washington, DC: The National Academies Press.

National Research Council. (2012a). *Climate Change Education in Formal Settings, K-14: A Workshop Summary*. A. Beatty, Rapporteur. Steering Commttee on Climate Change Education in Formal Settings, K-14. Board on Science Education, Division of Behavioral and Social Sciences and Education. Washington, DC: The National Academies Press.

National Research Council. (2012b). *Education for Life and Work: Developing Transferable Skills in the 21st Century*. Washington, DC: The National Academies Press.

National Research Council. (2013). *Next Generation Science Standards: For States, By States*. Washington, DC: The National Academies Press.

The New York Times. (2014). Environment. Available: http://green.blogs.nytimes.com [February 2014].

Nordhaus, W.D. (2012). Why the global warming skeptics are wrong. *New York Review of Books, 59*(5). Available: http://www.nybooks.com/articles/archives/2012/mar/22/why-global-warming-skeptics-are-wrong/?insrc=toc [February 2014].

Nordhaus, W.D. (2013). *The Climate Casino: Risk, Uncertainty, and Economics for a Warming World*. New Haven, CT: Yale University Press.

O'Neill Packard, K., and Reinhardt, F. (2000). What every executive needs to know about global warming. *Harvard Business Review, 78*(4), 128-135.

Patenaude, G. (2010). Climate class for business schools. *Nature, 466*(7302), 30.

Patenaude, G. (2011). Climate change diffusion: While the world tips, business schools lag. *Global Environmental Change, 21*(1), 259-271.

Pielke, R. Jr. (2010). *The Climate Fix*. Philadelphia, PA: Perseus Books.

Pielke, R. Jr. (2014). Blog. Available: http://rogerpielkejr.blogspot.com [February 2014].

Reinhardt, F.L., and Hyman, M. (2007, revised 2009). *Global Climate Change and BP*. Harvard Business School Case 708-026. Available: http://www.hbs.edu/faculty/Pages/item.aspx?num=34984 [February 2014].

Reinhardt, F.L., and O'Neill Packard, K. (2001). A business manager's approach to climate change. In E. Claussen (Ed.), *Climate Change: Science, Strategies, & Solutions*. Leiden: Brill Academic.

Shattuck, R. (2009). *A Closer Look at Business Education: Sustainability in the Curriculum*. Report for the Aspen Institute Center for Business Education. Available: http://www.aspencbe.org/documents/SustainabilityintheCurriculum.pdf [February 2014].

Smith, L.C. (2011). *The World in 2050: The Four Forces Shaping Civilization's Northern Future*. New York: Plume.

Stott, P.A., Gillett, N.P., Hegert, G.C., Karoly, D.J., Stone, D.A., Zhang, X, and Zwiers, F. (2010). Detection and attribution of climate change: A regional perspective. *Wiley Interdisciplinary Reviews: Climate Change, 1*(2), 192-211.

Sundaram, A. (2009). What is the link between business and climate change? Blog post. Available: http://tuckatcop15.wordpress.com/2009/12/09/what-is-the-link-between-businesses-and-climate-change-by-professor-anant-sundaram/ [February 2014].

Sundaram, A. (2012). What is a business school doing in Doha, at COP18? Blog post. Available: http://tuckcop18.wordpress.com/2012/11/26/whats-a-business-school-doing-in-doha-at-cop18-professor-anant-sundaram/ [February 2014].

Sundaram, A., Karras, C., and Mengelt, C. (2013). Climate Education in Business Schools: A Brief Survey of the Landscape. Presentation at the Workshop on Climate Change Education for Future Business Leaders, March 14, National Academy of Sciences, Washington, DC.

Tol, R.S.J. (2009). The economic effects of climate change. *Journal of Economic Perspectives, 23*(2), 29-51.

Trucost. (2011). *Universal Ownership: Why Environmental Externalities Matter to Institutional Investors*. Available: http://www.unpri.org/viewer/?file=files/6728_ES_report_environmental_externalities.pdf [February 2014].

U.S. Climate Action Partnership. (2009). *A Blueprint for Legislative Action: Consensus Recommendations for U.S. Climate Protection Legislation*. Available: http://www.us-cap.org/pdf/USCAP_Blueprint.pdf [February 2014].

The Wall Street Journal. (2012). Economics: Creating environmental capital. Available: http://economics.wsj.com [February 2014].

Whittle, P., and Campbell, D. (2014). Climate debate daily. Available: http://climatedebatedaily.com [February 2014].

Winn, M., Kirchgeorg, M., Griffiths, A., Linnenluecke, M.K., and Gunther, E. (2011). Impacts from climate change on organizations: A conceptual foundation. *Business Strategy and the Environment, 20*(3), 157-173.

Wittneben, B.B.F., and Kiyar, D. (2009). Climate change basics for managers. *Management Decision, 47*(7), 1122-1132.

The World Bank. (2012). *Turn Down the Heat: Why a 4°C Warmer World Must Be Avoided.* Available: http://climatechange.worldbank.org/sites/default/files/Turn_Down_the_heat_Why_a_4_dgree_centrigrade_warmer_world_must_be_avoided.pdf [February 2014].

Appendix A

The Climate Change Education Roundtable

Climate Change Education (CCE) is becoming a major investment area for the federal government and will involve, among others, the National Science Foundation (NSF), the National Oceanic and Atmospheric Administration (NOAA), the National Aeronautics and Space Administration (NASA), the Environmental Protection Agency (EPA), the U.S. Geological Survey (USGS), the U.S. Department of Education, and the U.S. Department of Energy. CCE is defined as an effort that seeks to ensure that individuals and communities understand the essential principles of Earth's climate system and the impacts of climate change and are able to make informed and responsible decisions with regard to actions that may affect climate and adaptation to climate change. Such broad goals for CCE demand a transdisciplinary approach that blends education, learning, social, behavioral, economic, and global Earth system science and requires careful consideration of related research evidence from each of these disciplines.

The National Research Council established the Roundtable on Climate Change Education in September 2009 to foster ongoing discussion of the challenges to and strategies for improving public understanding of climate science and climate change among federal agencies, regional and local government units, the business community, and nonprofit and academic sectors. Through its five meetings and two associated workshops over a course of approximately three years, the 30 Roundtable members were learning from each other and invited experts about issues such as the challenges inherent to climate change education; strategic approaches to designing interlocking programs in research, learning, workforce development, and public literacy; and strategies for accelerating the education of U.S. citizens about climate and climate change. The Roundtable brings together federal and state policy makers, educators, scientists and communications and media experts. It includes a number of *ex officio* members from federal agencies with dedicated interests in climate change education, including officials from NSF, NOAA, NASA, USGS, the Department of Education, the Department of Energy, the Department of Agriculture, and EPA.

The CCE Roundtable provides an opportunity to bring together overlapping and complementary expertise from academic and professional disciplines that commonly do not intersect when addressing CCE. It also provides federal agencies with important foundational knowledge related to key aspects of CCE and learning, such as the nature and scope of existing efforts, achievable and measurable goals, challenges and opportunities inherent in developing a national-level CCE initiative, and areas where investments may provide the greatest leverage. Roundtable discussions also provide useful new insights for a variety of other stakeholders. The CCE Roundtable also provides a formal mechanism to support continued collaboration and cooperation across federal agencies on major future climate change education or other science education initiatives. Through Roundtable discussions, the work of the federal agencies can be coordinated with stakeholders from private and nonprofit sectors such that their efforts can be built to complement and enhance federal initiatives. Through its public workshops and published workshop summaries, the CCE Roundtable will also become a source for evidence-based information related to climate change education.

So far, the CCE Roundtable has sponsored four associated workshops. The first one focused on fundamental goals and objectives of climate change education and outreach to various target audiences, and ways to understand whether these goals and objectives are being met. A summary report for the workshop is available at http://www.nap.edu: *Climate Change Education: Goals, Audiences, and Strategies: A Workshop Summary* (National Research Council, 2011, Washington, DC: The National Academies Press).

The second workshop sponsored by the Roundtable focused on K-14 education, particularly in light of the NRC conceptual framework for new science education standards that included climate change education-relevant aspects. A summary report for the workshop is available at http://www.nap.edu: *Climate Change Education in Formal Settings, K-14: A Workshop Summary* (National Research Council, 2012, Washington, DC: The National Academies Press, 2012).

The workshop on climate change education for future business leaders constituted the third Roundtable-sponsored workshop. The fourth workshop discussed strategies that are rooted in behavioral, social, and educational sciences for engaging private individual, family, and community forestland owners in conversations about climate and climate change. The workshop was held in August 2013, and a report is forthcoming.

Appendix B

Workshop Agenda

The National Academies
Climate Change Education Roundtable

A Workshop on
Climate Change Education for Future Business Leaders

Thursday, March 14, 2013
National Academy of Sciences
2101 Constitution Avenue
Washington, D.C.

Objectives:

1. Discuss climate change as a key strategic business issue with a focus on risk and opportunities.
2. Learn from leaders in business schools and international corporations how they train the necessary talent.
3. Identify the institutional barriers and discuss models that work for incorporating climate change education into business schools.
4. Discuss pathways to providing climate change education for future business leaders.

March 14, 2013 (Room 120)

Breakfast available at the meeting room starting at 7:30 a.m.

8:00 a.m. **Welcome & Workshop Goals**
Janet Peace, *VP of Markets and Business Strategy, C2ES*
Martin Storksdieck, *Director, Climate Change Education Roundtable*

8:15 a.m. **Climate Change: The Global Challenge**
Robert W. Corell, *Global Environment and Technology Foundation*

9:00 a.m. **Business Opportunities and Challenges Posed by Climate Change**
Moderator: Janet Peace

Panelists:
Dave Nagel, *Executive VP, BP America Inc.*
Michael Parr, *Senior Manager, Government Affairs, DuPont*
Valerie Patrick, *Sustainability Coordinator, Bayer*
Camilla Seth, *Executive Director, Environmental Affairs, JPMorgan*

10:30 a.m. *Break*

10:45 a.m. – 12:15 p.m. **Business Schools: Opportunities and Challenges Posed by Climate Change**
Moderator: Richard Vietor, *Senior Associate Dean for the Asian Initiative, Harvard Graduate School of Business Administration*

Panelists:
Robert Hansen, *Senior Associate Dean, Tuck School of Business at Dartmouth*
Valerie Suslow, *Associate Dean, Ross School of Business, University of Michigan*
Anjani Jain, *Senior Associate Dean, School of Management, Yale*

12:15 p.m. *Lunch (provided onsite)*

1:15 p.m. **Current Efforts and Approaches for Incorporating Climate Change Education into Business Schools**
Introduction by Anant Sundaram, *Professor of Finance, Tuck School of Business, Dartmouth College*

Moderator: Andy Hoffman, *Holcim Professor of Sustainable Enterprise, University of Michigan*

Panelists:
Eric Orts, *Director, Initiative for Global Environmental Leadership, Wharton*
Jacob Park, *Associate Professor of Business Strategy and Sustainability, Green Mountain College*

Dan Reicher, *Executive Director, Steyer-Taylor Center for Energy Policy and Finance, Stanford University*
Melissa Paschall, *Director, Business & Environment Initiative, Harvard Business School*

3:15 p.m. *Break*

3:30 p.m. **Needs and Opportunities: A Way Forward**
Moderator: Anant Sundaram

Discussants:
Laura Clise, *Director of Sustainable Development, AREVA Inc.*
Bruce Schlein, *Director of Corporate Sustainability, Citi*
Dan Vermeer, *Executive Director, Center for Energy, Development and the Global Environment, Duke University*

Attendees and Discussants will reflect on the needs, opportunities, and ways forward.

4:30 p.m. **Review of Main Ideas**
Janet Peace

4:45 p.m. **Closing Remarks**
Martin Storksdieck

5:00 p.m. **Networking Event**

6:00 p.m. **Meeting Adjourns**

Appendix C

Registered Workshop Participants

Rae Andre (Northeastern University)
Tina Bailey (Queen's School of Business)
Marcia Barton (National Science Foundation)
James Brey (American Meteorological Society)
Anthony Brown (National Research Council)
Judsen Bruzgul (American Meteorological Society)
David Campbell (National Science Foundation)
Lin Chambers (NASA Langley Research Center)
William Chernicoff (Toyota Motor North America, Inc.)
Laura Clise (AREVA Inc.)
Robert Corell (Global Environment and Technology Foundation)
Tina Dacin (Queen's School of Business)
Rebecca Davis (Metropolitan Washington Council of Governments—Clean Air Partners)
Laura Delgado-López (Institute for Global Environmental Strategies)
Jenny Dissen (National Oceanic and Atmospheric Administration)
Glen Dowell (Cornell University)
Edward Dunlea (National Academy of Sciences)
Alexander Dymersky
Lynn Elfner (The Ohio Academy of Science)
Aimee Dars Ellis (Ithaca College)
Evelina Felicite-Maurice (National Aeronautics and Space Administration)
Maya Fischhoff (Network for Business Sustainability)

Paul Fleming (Seattle Public Utilities)
Sherrie Forrest (National Research Council)
Laurie Geller (National Research Council)
Bradley George (Babson College)
Jim Geringer (ESRI)
Marcus Griswold (University of Maryland)
Robert Hansen (Dartmouth College)
Elizabeth Harball (Environment and Energy Publishing)
Stefan Heck (McKinsey & Company)
Rebecca Henderson (Harvard University)
Andrew Hoffman (University of Michigan)
DaNel Hogan (U.S. Department of Energy)
Steve Hunt (greendefenseinitiative.com)
Susan Jackson (Rutgers University)
Anjani Jain (Yale University)
Jasper Jones (UBCA Community Trust, Inc.)
Paige Kelly (Boston College)
Neil Leary (Dickinson College)
Tamara Ledley (TERC)
Charles Lieder (Shell)
Mark McCaffrey (National Center for Science Education)
Claudia Mengelt (National Research Council)
Silva Meybatyan (University of the District of Columbia)
Jerry Miller
Elizabeth Mills (American Meteorological Society)
David Nagel (BP America, Inc.)
Frank Niepold (National Aeronautics and Space Administration)
Anastasia Northland (University of Miami)
Mary Oleksy (Stanford University)
Eric Orts (University of Pennsylvania)
Jacob Park (Green Mountain College)
Michael Parr (DuPont)
Doug Parsons (National Park Service)
Jill Parsons (Ecological Society of America)
Melissa Paschall (Harvard University)
Sanjay Patnaik (University of Pennsylvania)
Valerie Patrick (Bayer)
Michael Pawlish (Montclair State University)
Janet Peace (Center for Climate and Energy Solutions)
Valerie Peckham (Philadelphia Zoo)
William H. Prado (Green Mountain College)
Mark Proegler (BP America, Inc.)
Dan Reicher (Stanford University)

Liesl Anna Riddle (George Washington University)
Jamesine Rogers
Katherine Rowan (George Mason University)
Bruce Schlein (Citi)
Theresa Schwerin (Institute for Global Environmental Strategies)
Camilla Seth (J.P. Morgan)
Martin Storksdieck (National Research Council)
Anant Sundaram (Dartmouth College)
Valerie Suslow (University of Michigan)
David Szymanski (Bentley University)
Galen Treuer (University of Miami)
Louie Tupas (National Institute of Food and Agriculture)
Marc Ventresca (University of Oxford)
Dan Vermeer (Duke University)
Richard Vietor (Harvard University)
Ming-Ying Wei (National Aeronautics and Space Administration)
Phil Weilerstein (National Collegiate Inventors and Innovators Alliance)
Doug Young (Broward County)

Appendix D

Biographical Sketches of Steering Committee and Speakers

STEERING COMMITTEE

JANET PEACE (*Chair*) is the vice president of markets and business strategy at the Center for Climate and Energy Solutions. In this role, she manages the center's Business Environmental Leadership Council. In addition, she manages the center's economics program and its analysis of market-based policy options. Previously, she held the same role at the Pew Center on Global Climate Change, and she was director of Offsets Development and Industry Relations at the Canadian nonprofit C3 (formerly Climate Change Center). She has taught environmental and natural resource economics at the University of Calgary, worked as a resource specialist with the U.S. General Accounting Office, and as a geologist with the U.S. Geological Survey. She is an expert member of the Climate Change Education Roundtable.

STEFAN HECK is director of sustainability and resource productivity at McKinsey & Company. He leads McKinsey's work in clean technology. Prior to joining McKinsey, he helped develop Apple's web presence; built a prototype satellite image-delivery system for NASA; founded the Stanford Solar Car Project, building two generations of solar-powered cars; developed expert systems for IBM; and developed pattern-recognition neural networks at Xerox PARC. He previously led McKinsey's Semiconductor Practice.

ANDY HOFFMAN is the Holcim (U.S.) professor of sustainable enterprise at the University of Michigan, a position that holds joint appointments at the Stephen M. Ross School of Business and the School of Natural Resources & Environment. Within this role, he serves as director of the Frederick A. and Barbara M. Erb Institute for Global Sustainable Enterprise. His research uses a sociological perspective to understand the cultural and institutional aspects of environmental issues for organizations. He was a member of the National Research Council Panel on Addressing Climate Change Through the Behavioral and Social Sciences.

MARK PROEGLER is director of climate and transport energy policy at British Petroleum (BP). Prior to joining the Climate and Transport Energy Policy group, he worked in Australia with BP's Corporate Environmental Policy team. He has also held a number of managerial, operational, marketing, and business development positions with BP and the former Amoco Corporation, including gas and power, chemicals, and marketing and refining.

BRUCE SCHLEIN joined Citigroup in 2006, where he now works as director of corporate sustainability. He leads an internal working group on energy efficiency financing. Previously, he worked as a sustainability specialist for Bechtel Corporation on petrochemical and civil projects in China and Romania, and for international development agencies including Save the Children, Catholic Relief Services in Bosnia-Herzegovina, and the U.S. Peace Corps in Papua New Guinea.

ANANT SUNDARAM is on the finance faculty at Tuck School of Business, Dartmouth College. His areas of expertise are business valuation, mergers and acquisitions, corporate governance, and financial strategies for profitable growth. He has published widely in law, finance, and management journals, as well as in the popular press. Sundaram pioneered numerous MBA and executive education courses, including the first course on business and climate change at a leading U.S. business school. He created the Fossil Fuel Beta (FFß), a metric to measure the stock price impact of a company's exposure to fossil fuel price changes and CO_2 emission risks.

RICHARD VIETOR is the Paul Whiton Cherington professor of business administration at the Harvard Graduate School of Business Administration, and senior associate dean for the Asian Initiative. Before joining the business school in 1978, he held faculty appointments at Virginia Polytechnic Institute and the University of Missouri at Columbia. His research on business and government policy has been published in numerous journals and books. For his courses in business-government relations and environmental management, he has published more than 80 case studies. He has been a consultant to the Hudson Institute, the Energy Research and Development Administration, several corporations, and the Prime Minister of Malaysia.

WORKSHOP SPEAKERS

LAURA CLISE is the director of external communications at AREVA, responsible for external communications, stakeholder engagement, and corporate citizenship. She previously led AREVA North America's Sustainable Development department. Before joining AREVA, her sustainability experience included work for the U.N. Global Compact, Intel Corporation, and Business for Social Responsibility. She began her career in brand management at General Mills and subsequently worked for McMaster-Carr Supply Company, where she led the international sales operations department.

ROBERT W. CORELL is a principal at the Global Environment Technology Foundation and leads its Center for Energy and Climate Solutions. He was appointed vice chancellor of the Academy of Science and Arts in 2012. He is founder and chair of the Global Climate Action Initiative and, in 2010, founded the nonprofit Global Science Associates. He led the Arctic Climate Impact Assessment (2005) and a comprehensive study of governance issues in the circumpolar Arctic. He is actively engaged in research concerned with the sciences of global change and the interface between science and public policy.

ROBERT HANSEN is senior associate dean and Norman W. Martin 1925 professor of business administration at the Tuck School. He serves as faculty director of the Center for Business and Society and co-faculty director of the Business Bridge Program and the Master of Health Care Delivery Science Program at the college. His expertise is in microeconomics, with a focus on game theory, auction markets, and public policy. His teaching has been in the economics, public policy, and finance areas. He has written and published articles in the areas of auction theory, corporate finance and corporate policy, industrial organization, and law and economics.

ANJANI JAIN is senior associate dean at the Yale School of Management. Previously, he joined the faculty of the Wharton School of the University of Pennsylvania in 1986, where he served for 26 years. From 2000 to 2010, he served as vice dean and director of Wharton's Graduate Division and from 2010 to 2012 as vice dean of Wharton's MBA Program for Executives.

He has published in the professional literature on telecommunications network design, and his more recent work has investigated the impact of growing product variety on the manufacture and design of families of products, especially in the automobile industry.

DAVID NAGEL has over 30 years' experience in the global energy industry, with executive positions in international business, corporate finance, and government relations. Since July 2009, he has managed BP's engagement with Washington stakeholders. His career began with Amoco International in Chicago, and he joined BP following the merger with Amoco in 1999. In 2012, Nagel and three faculty members from Duke Corporate Education co-authored *The New CFOs: How Finance Teams and Their Leaders Can Revolutionize Modern Business*.

ERIC ORTS is the Guardsmark professor at the Wharton School of the University of Pennsylvania. He is a professor in the Legal Studies and Business Ethics Department with a joint appointment in the Management Department, and he directs the Initiative for Global Environmental Leadership (IGEL) at Penn/Wharton. He serves also as an academic co-director of the certificate program for securities compliance and regulatory professionals. Prior to joining Wharton's faculty in 1991, he practiced law at Paul, Weiss, Rifkind, Wharton & Garrison in New York City and was a Chemical Bank fellow in corporate social responsibility at Columbia Law School.

JACOB PARK is associate professor of business strategy and sustainability at Green Mountain College in Vermont, specializing in the business of social and environmental innovation and entrepreneurship. He serves on the Renewable Energy and Adaptation to Climate Technologies Investment Sub-Committee of the Africa Enterprise Challenge Fund and is the chair of the U.S. Sustainable and Responsible Investment Forum's International Working Group Steering Committee.

MICHAEL PARR is senior manager of government affairs at DuPont, where he manages activities related to science, technology, sustainability, environment, bio-based materials, energy, and climate issues for DuPont. His responsibilities include managing these issues with the executive and legislative branches of the federal government, interaction with nongovernmental organizations and other external stakeholders, and contributing to DuPont's global energy, environmental, and science policy process.

MELISSA PASCHALL is the director of the Business and Environment Initiative at Harvard Business School. Her experience includes sustainability consulting for large companies and nonprofits, case writing on topics related to corporate sustainability and mission-driven organizations, and the development of a climate change curriculum used by business students in several schools in Europe.

VALERIE PATRICK is sustainability coordinator for Bayer Corporation and Bayer Material Science, LLC. She develops strategy, and facilitates and tracks progress toward the focal points for Bayer's regional North American strategy on sustainable development. She serves as a board member of the Association of Climate Change Officers.

DAN REICHER is executive director of the Steyer-Taylor Center for Energy Policy and Finance at Stanford University. He has served three U.S. presidents, including as an assistant secretary of energy during the Clinton administration and as a member of President Obama's transition team. He is a member of the National Academy of Sciences Board on Energy and Environmental Systems and co-chairman of the board of the American Council on Renewable Energy.

CAMILLA SETH serves as executive director in JPMorgan Chase's Office of Environmental Affairs. She has advised leading commercial and investment banks on sustainability strategy formulation, environmental and social risk management, identification of environmentally superior investment opportunities, and stakeholder engagement. She has also worked in philanthropy for Citigroup and the Surdna Foundation. She was the founding director of Programs and Operations at the Global Impact Investing Network.

VALERIE SUSLOW is associate dean for graduate programs, professor of business economics and public policy, and Louis and Myrtle Moskowitz research professor of business and law at the Stephen M. Ross School of Business at the University of Michigan. She joined the faculty in 1984 and served as faculty director of the Evening MBA Program from 1995-1999 and associate dean for degree programs from 2007 to 2010. Her research focuses on the economics of price fixing, the duration and organizational features of cartels, the impact of international cartels on developing countries and on international trade more generally, and related antitrust policies.

DANIEL VERMEER directs Duke University's Center for Energy, Development, and the Global Environment. In this role, he teaches graduate classes at the Fuqua School of Business and Nicholas School of the Environment, directs research projects, and consults with leading companies and organizations. He is also a lead partner in Duke's Corporate Sustainability Initiative. Vermeer is the founder and chief architect of the Global Water Challenge, a multipartner organization for innovative water and sanitation initiatives, coauthor of the CEO Water Mandate, and lead contributor to several policy documents.